Heath Hen

by Joyce Markovics

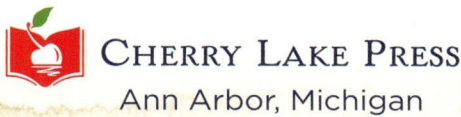

Ann Arbor, Michigan

Published in the United States of America by Cherry Lake Publishing Group

Ann Arbor, Michigan

www.cherrylakepublishing.com

Reading Adviser: Beth Walker Gambro, MS Ed., Reading Consultant, Yorkville, IL

Content Adviser: David R. Sischo, PhD, Wildlife Biologist

Book Designer: Ed Morgan

Photo Credits: Wikimedia Commons, cover and title page; Wikimedia Commons, 4; Wikimedia Commons, 5; Courtesy of the George J. Mitchell Department of Special Collections & Archives, Bowdoin College Library, Brunswick, Maine, 6; Wikimedia Commons, 7 top; Courtesy of the George J. Mitchell Department of Special Collections & Archives, Bowdoin College Library, Brunswick, Maine, 7 bottom; © Coatesy/Shutterstock, 8; © Maximillian cabinet/Shutterstock, 9; © Steve Oehlenschlager/Shutterstock, 10; © Nattapong Assalee/Shutterstock, 11; © Brian A Wolf/Shutterstock, 12; © Agami Photo Agency/Shutterstock, 13; Wikimedia Commons, 14; © Everett Collection/Shutterstock, 15; Wikimedia Commons, 16; Wikimedia Commons, 17; Wikimedia Commons, 18; Courtesy of the George J. Mitchell Department of Special Collections & Archives, Bowdoin College Library, Brunswick, Maine, 19; © Danita Delmont/Shutterstock, 20–21; Wikimedia Commons, 22.

Copyright © 2023 by Cherry Lake Publishing Group

All rights reserved. No part of this book may be reproduced or utilized in any form or by any means without written permission from the publisher.

Cherry Lake Press is an imprint of Cherry Lake Publishing Group.

Library of Congress Cataloging-in-Publication Data

Names: Markovics, Joyce L., author.
Title: Heath hen / by Joyce Markovics.
Description: Ann Arbor, Michigan : Cherry Lake Publishing, [2023] | Series: Endlings. The last species | Includes bibliographical references and index. | Audience: Grades 4-6
Identifiers: LCCN 2022004222 (print) | LCCN 2022004223 (ebook) | ISBN 9781668909652 (hardcover) | ISBN 9781668911259 (paperback) | ISBN 9781668912843 (ebook) | ISBN 9781668914434 (pdf)
Subjects: LCSH: Greater prairie chicken—Juvenile literature. | Extinct animals—Juvenile literature.
Classification: LCC QL696.G27 M36 2023 (print) | LCC QL696.G27 (ebook) | DDC 598.6/2–dc23/eng/20220329
LC record available at https://lccn.loc.gov/2022004222
LC ebook record available at https://lccn.loc.gov/2022004223

Printed in the United States of America by Corporate Graphics

THE LAST SPECIES CONTENTS

Booming Ben 4
Heath Hens 8
Facing Extinction 14
Bringing Ben Back 20

Animals Under Threat 22
Glossary 23
Find Out More 24
Index 24
About the Author 24

BOOMING Ben

Tucked in the woods in Massachusetts is a large statue. However, it doesn't show a famous person. It's a **memorial** to a bird. The bird's name is Booming Ben. And he was the last heath hen that ever lived. The last known animal of its kind is called an endling.

The statue of Booming Ben is in Martha's Vineyard, an island off the coast of Massachusetts.

The statue stands in the area where Ben once roamed. He earned his name after the booming sounds he made to attract a **mate**. However, when Ben died, there were no female—or any other—heath hens left on Earth.

Artwork showing a male and a female heath hen

Heath hens were large, chicken-like birds.

Scientist Alfred O. Gross studied and watched Ben before he died. "The bird presented a **pathetic** figure as it stood out there all alone," he wrote. Booming Ben's only companions were "the crows that had come to share [his] food."

Scientist Alfred O. Gross

Alfred and some other scientists talked about capturing Ben to keep him safe. Alfred felt that Ben should live out his days as a free bird. The last heath hen will "die in the surroundings in which it has lived," Alfred said. And that, sadly, is what happened in 1932. No one knows exactly how Booming Ben died.

A stuffed female heath hen at a museum in Boston

Alfred holds Ben after he died.

Alfred attached metal bands to Ben's legs. He wanted to be able to identify his remains if a predator, such as a hawk, caught him.

Heath Hens

Because they've been **extinct** for decades, heath hens are not well-known birds. They are a type of grouse. Grouse are smallish birds with plump bodies, round wings, and short tails. Prairie chickens, which are not chickens at all, also belong to the grouse family. And heath hens are one of two types of prairie chicken.

There are six types of grouse that live in North America.

This is the other kind of prairie chicken. It's called the greater prairie chicken.

Before they went extinct, heath hens lived in tallgrass **prairies**. They ate seeds, leaves, and small fruits such as berries. In summer, when insects are plentiful, they ate crickets, flies, and beetles.

Most birds in the grouse family are known as game birds. Why? They're often hunted for food or sport.

Male and female heath hens looked different from each other. Adult males had an orange band of feathers over their eyes. On their heads were long feather **tufts** that they could raise or lower. Most distinctive of all were their bright orange throat sacs. Males puffed up their throat sacs during mating season.

The orange eye combs, feather tufts, and throat sac of a male prairie chicken would have looked similar to a male heath hen.

In spring, males gathered on "booming grounds." These areas were where male heath hens tried to attract females. To do this, they lowered their heads, raised their tufts, and snapped their tail feathers. The males also made low, booming sounds using their throat sacs.

Male prairie chickens dancing and booming

As the male heath hens boomed, they often leaped into the air. Some cackled. Others stamped their feet. These displays **lured** female hens. Eventually, females mated with the best boomers.

A male and female prairie chicken

Female heath hens lined their nests with grass, leaves, and feathers.

After mating, the females built nests in tall grasses. They laid between 5 and 17 eggs. After about 3 weeks, the eggs hatched. At 10 weeks old, the little heath hens were ready to live on their own.

Female heath hens raised the young.

Facing Extinction

Heath hens once roamed from Virginia to New Hampshire. They were very common when settlers first came to North America. Some experts believe that Native Americans and Pilgrims ate heath hen—not turkey—for the first Thanksgiving! The birds were so common and easy to get, early settlers called them "poor man's food."

Early settlers farmed the land and hunted for food.

Because of their popularity, people overhunted heath hens. By the late 1700s, their numbers had sharply fallen. Incredibly, early Americans took action to save heath hens. This was the first time in history Americans tried to save a wild animal!

As early as 1791, a law was passed in New York. It was a bill "for the **preservation** of heath-hen and other game." However, the law was hard to **enforce**. Over the next decades, heath hens continued to be hunted.

Americans were also moving onto the prairies where heath hens lived. By 1870, there were no more heath hens left on the mainland United States. Only a small **population** remained on an island named Martha's Vineyard off the coast of Massachusetts.

The shores of Martha's Vineyard

By the late 1800s, there were only about 100 heath hens left.

Concerned scientists set up a heath hen **preserve** on Martha's Vineyard. By the early 1900s, the population had grown to about 2,000 birds. Then, in 1916, a fire swept across the preserve. It killed most of the heath hens, including almost all of the females and young.

A photo of a heath hen taken in the early 1900s

By 1927, only 13 heath hens remained. That number dropped to 2 in 1928. By the spring of 1929, there was only one heath hen left—Booming Ben. Scientist Alfred O. Gross looked after Ben until the bird died in 1932.

Scientist Alfred O. Gross holding a heath hen in Martha's Vineyard

Bringing Ben Back

The boom of a heath hen might once again be heard on Martha's Vineyard. An organization called Revive & Restore is trying to bring back the birds. How? Put simply, the group gets **DNA** from a dead animal. It uses that DNA and a **domestic** chicken to create a baby heath hen.

If the project moves forward, it would take 2 years and about half a million dollars. But many people are excited about the possibility of bringing Ben back. However, any new heath hens will need a place to live and roam. As one **conservation biologist** says, "If the habitat doesn't exist, the species won't persist."

Animals Under Threat

Many more birds are at risk of dying out. Here are three on the brink of extinction:

California Condor
It's the largest land bird in North America, and it feeds on dead animals. Around 500 of these birds are left.

Giant Ibis
This giant wading bird lives in Southeast Asia. Only around 200 exist.

New Caledonian Owlet-Nightjar
It's thought this wide-mouthed bird feeds on insects on an island in the South Pacific. Fewer than 50 remain.

Glossary

biologist (bye-OL-uh-jist) a scientist who studies living things

conservation (kon-sur-VAY-shuhn) the protection of wildlife and natural resources

DNA (DEE EN AY) the molecule that carries the genetic code for a living thing

domestic (duh-MESS-tik) describes animals that have been bred and tamed by humans

enforce (en-FORSS) to make sure that a law is obeyed

extinct (ek-STINGKT) when a kind of plant or animal has died out completely

lured (LOORD) tempted to come closer

mate (MATE) one of a pair of animals that have young together

memorial (muh-MOR-ee-uhl) something that is built to remember an animal, person, or event

pathetic (puh-THET-ik) sad or pitiful

population (pop-yuh-LAY-shuhn) the number of animals living in a place

prairies (PRAIR-eez) large areas of flat land covered with grass

predator (PRED-uh-tur) an animal that hunts other animals for food

preservation (preh-zuhr-VAY-shuhn) the act of keeping something in existence

preserve (pri-ZURV) a place where animals are kept safe and protected

tufts (TUFTS) small clumps

Find Out More

Books

Hoare, Ben, and Tom Jackson. *Endangered Animals*. New York, NY: DK Children, 2010.

Riera, Lucas. *Extinct: An Illustrated Exploration of Animals That Have Disappeared*. New York, NY: Phaidon Press, 2019.

Whitfield, John. *Lost Animals*. New York, NY: Welbeck Publishing, 2020.

Websites

The Bryan Museum: The Lost Bird Project
http://www.lostbird.org

National Geographic: The Photo Ark
https://www.nationalgeographic.org/projects/photo-ark/

Six Extinctions: An Overview of the Ends of Species
https://www.amnh.org/shelf-life/six-extinctions

Index

biologist, 21
California condor, 22
conservation, 19, 21
extinction, 8–9, 14, 22
giant ibis, 22
Gross, Alfred O., 6–7, 19
grouse, 8–9
heath hens
  Booming Ben, 4–7, 11, 19
  diet, 9
  feathers, 10–11, 13
  females, 5, 7, 10–13, 18

hunting, 9, 15–16
males, 10–11
mating, 10, 13
nests, 13
laws, 17
Massachusetts, 4, 17
Native Americans, 14
New Caledonian owlet-nightjar, 22
Pilgrims, 14
prairie chicken, 8–9, 11–12
preservation, 16

About the Author

Joyce Markovics has written hundreds of books for kids. She hopes this book inspires young readers to learn more about endangered animals and take action to prevent their extinction. She dedicates this book to the scientists who help wildlife survive and stay wild.